WhoA!

NATIONAL GEOGRAPHIC
KiDS

WeiRdeST
ANiMALS
ON THE PLANET

NATIONAL GEOGRAPHIC
WASHINGTON, D.C.

Octopuses have nine brains.

4

BLUE-RINGED OCTOPUS

This magnificent mollusk is only about the size of a golf ball, but it's big on weirdness. Like many octopuses, it's a camouflage pro. Octopuses have special cells, called chromatophores, that let them change color. When frightened, the blue-ringed octopus takes its color changing to the next level. It's like something straight out of a sci-fi movie: Shiny blue rings light up all over the octopus's body. This stunning show is really a dire warning. If it continues to feel threatened, the octopus can grab on with a nearly painless bite and deliver an extremely powerful venom ... enough to kill 20 people in just a few minutes!

Octopuses don't have teeth—they have beaks!

SEA OTTER

You *otter* know that these sea-dwelling mammals are more than just cute—they're also a little bizarre. When snoozing on their backs on top of the water, otters often will hold paws with each other so they don't get separated by the ocean currents. Otters also use a special part of their armpits to carry bits of food and other objects! These flaps of skin are frequently used to store rocks that they use to smash open the shells of clams and crabs to eat. For humans, that would be like keeping a can opener in your armpit!

GIRAFFE-NECKED WEEVIL

It's no wonder this insect reminds people of the world's most famous long-necked mammal. The male weevil's neck is usually two to three times longer than the female's. Males use it to fight with each other in hopes of impressing a mate. Not to be out-weirded by the males, female weevils do something strange with their eggs. While in a tree, they lay each egg inside a rolled tube of leaf, allowing the egg to then fall safely to the ground. After hatching, the newborn weevil noshes on the leaf as it grows.

7

SUN BEAR

This bear isn't being rude—it's just showing off! Sun bears' exceptional tongues can be up to 18 inches (46 cm) long! They use that long licker to slurp up foods like honey and insects. Their marvelous mouths are useful for more than just chowing down. Sun bears also seem to communicate by making funny faces at one another! Scientists once thought only people and other primates communicated this way.

Any hungry fish that grabs on to this animal may find itself a bit choked up. If caught, the hagfish secretes extra-strong slime, and a lot of it—enough of it to fill a five-gallon (19-L) bucket in minutes. This slick stuff clogs the other fish's gills, making it hard for it to breathe. If that doesn't put off an attacker, the hagfish's soft skin is also baggy enough to allow the hagfish to twist and slither free of the predator's jaws. Talk about getting out of a sticky situation!

PLANT OR ANIMAL?

B lending into its surroundings can help an animal both hunt prey and avoid being eaten. For some species, looking like a plant part isn't weird—it's a matter of survival.

DEAD-LEAF BUTTERFLY

Anyone there? When its wings are closed, the dead-leaf butterfly looks dull and dry, causing birds and other predators to think it's no more than an unappetizing brown leaf.

THORN BUG

When it comes to blending in, this insect gets right to the point. Its sharp-looking body is usually enough to fool possible predators into thinking it's not an insect.

MOSSY LEAF-TAILED GECKO

The name says it all! These geckos camouflage so well against rough and mossy tree bark that they're almost impossible to spot.

WhoA!

PANDA

Could you imagine eating only one thing for the rest of your life ... and *liking* it? If you answered yes, you might be a giant panda. They're un-*bear*-ably picky about their diets, eating nothing but bamboo. This preference has puzzled some scientists because pandas seem to have evolved with the ability to digest meat, like other bears have. But there's not a lot of competition for bamboo—the plants are tough and not very nutritious—so over millions of years, pandas stopped eating meat and began eating just the green stuff ... a lot of it. To get the nutrients they need, pandas eat at least 26 pounds (12 kg) of it a day!

CHRISTMAS TREE WORM

These festive frills aren't holiday decor—they're the crown of an ocean-dwelling worm! Despite its showy appearance, the Christmas tree worm is really quite shy. Most of its wormy body is forever tucked away in a tubelike burrow in the coral where it lives. Only its flashy crown peeks out. The frills help the worm breathe and capture food. But if a predator comes by, the worm vanishes into its tube in the blink of an eye.

FLYING SQUIRREL

Whoosh! These adorable aerial acrobats soar through the sky thanks to the big flaps of skin that connect their legs and body. Technically, they aren't really flying—they're gliding! But that doesn't make them any less impressive: These small critters can soar between trees that are 500 feet (150 m) apart. (That's about one and a half football fields!) They can even turn halfway around (180 degrees) in the air to avoid being picked off by a hawk or other big bird.

GLASSWING BUTTERFLY

In a game of hide-and-seek, this insect would win every time. The glasswing butterfly's wings are partially see-through, making them nearly invisible whether flying in the air or resting on a leaf. The animal may look delicate, but it's not as fragile as it seems. During migration season, these butterflies can fly thousands of miles. Plus, they've got a dangerous side: Both the caterpillar and adult forms of the glasswing are toxic to predators.

RHINOCEROS HORNBILL

Sometimes people use the phrase "an odd duck" to describe something that's different, but "an odd hornbill" might be more appropriate. When it's time to lay eggs, some species of female hornbill build their nests in holes in tree trunks. Then the female traps herself inside the nest, with the help of her mate. The pair leave only a small space open, filling in the rest of the hole with chewed-up sticks, mud, food, and even poop. The female stays walled up inside the tree with her eggs, sticking her bill out to get food that her mate brings to her. It's odd, but it helps keep the eggs safe from predators like snakes, monkeys, and other birds.

These birds' curved horns, called casques, act like echo chambers, making their calls loud enough to be mistaken for a lion's roar.

A hornbill's "eyelashes" aren't hair—they're feathers.

LEAF SHEEP

The name leaf sheep may sound like an animal that belongs in a pasture, but these green sea slugs actually live in ocean waters near Japan and the Philippines. They feed on green algae that contain chloroplasts, which, once inside the leaf sheep's body, absorb light from the sun and help the slug make its own food. Leaf sheep are one of the few animals in the world known to have this weird power!

TIGER

When people think of tigers, they usually think of words like "wild" and "fierce," not "weird." Maybe they should. Why? For one thing, a tiger's pee has a chemical in it that makes it smell like buttered popcorn. Second, a tiger's famous stripes are also kind of odd: They're not just on its fur! A peek under that orange-and-black pelt shows that the stripy pattern is also on the tiger's skin. And that's not all—a tiger's stripes are like human fingerprints. Every tiger cub is born with a pattern that's completely its own.

So StranGe!

CASSOWARY

Attention! The cassowary looks ready for battle, thanks to the helmet-shaped structure on top of its head, called a casque. It amplifies the booming call of the cassowary, making it the bass guitar of the avian world—it has the lowest pitch of any bird on Earth! The birds also communicate with their weird-looking wattles. These colorful flaps of skin hanging from their necks probably give clues to other cassowaries about each bird's mood.

Even the best dentist might not know what to do with the babirusa's teeth. A pair of gnarly teeth on the top jaw of the male babirusa continue to grow throughout the animal's life. The teeth grow through the roof of the animal's mouth, poke up through its snout, keep extending toward the animal's forehead, and can even push into its skull. (Fortunately, this doesn't happen very often.) The tusks are used to fight other males and tend to break off in the process.

Orchid mantises are one of the only animals that look like a flower blossom!

22

ORCHID MANTIS

Weird? Pretty? Both? Whatever words you use to describe this insect, make sure you include "deadly." The body of a young female mantis, also known as the nymph, looks just like an orchid. Orchids are flowers that are rich in sweet nectar, which attracts insects looking for food. Any insect fooled by the mantis's shape can quickly go from looking for a meal to becoming one. Like all praying mantises, this one is a speedy hunter. It will snag anything that gets too close to its long arms—including other orchid mantises. Females are especially likely to eat their smaller male counterparts. Males grow to about an inch (2.5 cm) long while females can reach a length of more than 2.5 inches (7 cm).

Male orchid mantises don't look like flowers at all; they are green and brown.

ALBATROSS

No landlubbers here! Albatross are seriously sea-loving animals. The webbed feet of some young albatross don't even touch land until they're at least six years old. When they're not paddling in the sea, albatross take to the air. These giant birds can ride ocean breezes for hours without stopping to rest or even needing a flap of their enormous—up to 11 feet (3.5 m) across!—wings. Albatross can even take mini-naps while they fly by shutting down part of their brains midair.

PINK FAIRY ARMADILLO

Here's a *fairy* cool critter! At only about five inches (13 cm) in length, this enchanting mammal is the world's smallest species of armadillo. It's also the only one that's pink, courtesy of all the blood vessels just under the surface of its shell. This tiny armadillo has an oddly magical way to defend itself. When it spots danger, it uses its sharp claws to dig a hole in the sandy ground. Then, it dives in headfirst, using its shell as a plug. In just a matter of seconds—*poof!*—the armadillo disappears!

KOALA

Picture a jelly bean. Now, imagine a baby koala. Did you picture them the same size? Believe it or not, they are! A newborn koala is only about 0.8 inch (2 cm) long. These tiny mammals may be blind, hairless, and lacking the round, fuzzy ears of their parents, but they're not helpless. They use their keen senses of smell and touch to locate their mother's pouch. Then, they pull themselves into it. They'll stay there, safe and warm, for about seven months, until they're big enough to peek out at their forest home—in Australian eucalyptus trees.

newborn koala

It's not a *pig*-ment of your imagination, this is a real animal! Found only in a very small region of India, this purple frog, also known as the pignose frog, stays underground almost its entire life. It can burrow up to three feet (1 m) deep, living on the insects and termites that fit into its teeny-tiny mouth. If that's not weird enough, people who study these critters say that male frogs make a call that sounds a lot like a chicken's cluck—and that's no hogwash.

WEIRDEST NAMES

When a new animal is scientifically described, people sometimes choose names that are pretty ordinary. Other times? Not so much!

FLAMINGO TONGUE SNAIL

FRIED EGG JELLYFISH

YODA PURPURATA

SARCASTIC FRINGEHEAD

29

DUGONG

Once upon a time, a swimming dugong may have inspired tales of mermaids and other mythical sea creatures. Today, they've got a barnyard-inspired nickname—sea cows. It actually makes a lot of sense for these grazers, which gather in herds of hundreds to pluck plants from seagrass fields. In fact, dugongs are one of the few ocean-dwelling mammals that are herbivores. *Udderly* amazing!

RESPLENDENT QUETZAL

Holy guacamole! At first glance, this shimmering bird's diet of fruit, lizards, and insects might not seem too weird. Things get a little wacky, though, when it comes to the way a quetzal eats wild avocados. It swallows the avocado whole, then throws up the fruit's large single seed. It sounds disgusting, but it's great for the avocado tree population. If the seed falls in a place with enough water and sunlight, it can grow into a new tree.

HAIRY FROG

Animals do what it takes to survive in the wild. This amphibian will break its own toes (you read that correctly) and use the pointed ends of its bones to scare off predators—earning it the nickname "wolverine frog." As if that's not hair-raising enough, males also grow fur-like strands of skin and blood vessels around breeding time. These fuzzy fringes probably help them take in more oxygen from the water, so they can stay underwater for days—it's the male's job to guard the eggs, so this ability would be useful.

Road trip! When it's time to travel, carrier crabs often don't travel alone. These crustaceans are fond of snagging a nearby sea urchin and giving it a piggyback ride from place to place. It's a win-win for the animals. The sea urchin gets a free ride to a new place to look for food, and the crab gets a spiky shield that protects it from predators.

STAR-NOSED MOLE

This mammal's name may include the word "star," but its nose looks more like a squid's squirming tentacles than anything twinkling in the night sky. The fleshy parts of the mole's nose are like fingers that help it feel its way through underground tunnels as it hunts for worms and insects. When it finds its prey, mealtime is over in a flash! Star-nosed moles have been called the world's fastest eaters. They can grab and swallow their prey in less than half a second. Even stranger, star-nosed moles often live in marshy areas and are excellent swimmers and underwater hunters. They snort bubbles from their noses and then suck them back in, sniffing them to follow the scent of a meal.

Brrr! Scientists have observed these moles swimming in ponds under a layer of ice.

The moles use their front feet and claws like shovels.

35

TROLL DOLL BUG

Every day is a bad hair day for this insect. The bug gets its name from this terrific tuft, which reminded the scientists who named it of the iconic troll dolls. The teeny animal is only about 0.3 inch (7 mm) long, not including the hairy-looking tuft that sticks up and out of its rear. Scientists think that the hair might give the animal a chance to escape from predators. When a bird or snake grabs at the tuft, it breaks off, giving the bug a chance to escape from a hairy situation.

AARDVARK

Dig this creature! Aardvarks are such speedy diggers, they can scoop out a burrow in five minutes flat. They're also a bit fussy about their digs, making a new burrow to hide out in every day. The aardvark cautiously emerges from its hole around sunset to head off in search of a bug feast. When it finds a termite mound, it uses its bearlike claws to rip into it, then lowers its long tongue down inside. Insects get stuck to the aardvark's sticky spit ... and, *slurp,* it's chow time!

BEARDED VULTURE

No bones about it—this bird has some strange feeding habits. Like all vultures, the bearded vulture feeds on carrion, the remains of dead animals. However, most of its diet—between 70 and 90 percent—is made up of bones, which tend to be rejected by even the least picky of other vulture species. That's because bearded vultures have special digestive juices that can break down bone. These clever birds also know how to get the good stuff, called bone marrow, that's hidden inside the bone. They drop bones from up to 260 feet (80 m), shattering them on rocks below. *Bone* appétit!

Australia, New Zealand, and New Guinea are home to some of Earth's oddest animals. And you can definitely include the echidna on this list of strange critters! Why? For starters, it's one of the few mammals that lays eggs rather than giving birth to live young. Once its babies hatch, they're called puggles. Another standout trait: the echidna's body temperature, which is the lowest of all mammals, at around 89°F (32°C). Scientists think it's one of the reasons echidnas live so long—more than 40 years—compared with other small mammals.

MARY RIVER TURTLE

Green hair, don't care! The punk "hairstyle" on the one-of-a-kind Mary River turtle is actually made of algae. Its green, stringy appearance helps the animal hide at the bottom of the Mary River in Queensland, Australia. Its two unusual chin whiskers, called tubercles, help it feel its way around in the muck. And once the turtle is down there, it can stay underwater for up to 72 hours. Even cooler? Mary River turtles are one species in a group of turtles that can breathe through their rear ends. A special body part inside the turtle's backside helps it take oxygen from the water.

The Mary River turtle is endemic to the Mary River, meaning it is not found anywhere else on Earth.

These turtles don't reach adulthood until they're at least 25 years old.

LONG-HORNED ORB WEAVER

If you mess with this spider, you'll have to deal with its horns. Female long-horned orb weavers have a huge pair of inward curving spines. These striking spines can be more than an inch (26 mm) long—that's about three times as long as the width of the spider's body! The spines make this animal a less than tasty meal for predators, and one tough critter to swallow. Despite their fierce appearance, the spiders are pretty shy. Unlike many other spiders, these orb weavers don't hang out in the center of their webs. They prefer to hide nearby.

Looking for an icebreaker? Polar bears are known for their characteristic white fur, but the truth is that a polar bear's fur isn't actually white—it's see-through! We see the bear's coat as white because of the way light passes through the individual hairs that make up the fur. In fact, a polar bear's skin is actually black. Why? The dark coloring helps absorb sunlight, keeping the animal warm in the ice-cold Arctic.

WhoA!

SECRETARY BIRD

In a hunting competition, the secretary bird would have a real leg up on the other contestants. When this large bird spots a snake or other potential meal, it will sometimes stomp on it, overcoming it by using its clawed feet and strong legs. Its legs also account for a good bit of its impressive height. Secretary birds can be four to five feet (1.2 to 1.5 m) tall—about the size of an 11- to 13-year-old human. Although they can fly, the massive birds prefer to stay on the ground, hunting in pairs or small groups.

COCONUT OCTOPUS

One animal's trash is another animal's treasure. At least, that's true for the coconut octopus. When another creature eats the meat of a tasty coconut, this octopus salvages the leftover shell to use as protection. Even with eight limbs, carrying around empty coconut shells is a challenge. The octopus uses six of its legs to grab and hold the shells, and the other two to move. This makes the octopus look like it's walking on stilts! If a predator comes along, the cephalopod can slither under the shells and hide out until it's safe. Super smart!

45

WEIRD RELATIONSHIPS

Sometimes survival requires a little assistance from another animal. Here are a couple of ways animals interact that are just a bit odd.

OXPECKERS AND GIRAFFES

Oxpeckers pull pesky insects from the fur of large animals, like giraffes. Unfortunately for the giraffes, the birds wash those insects down with sips of the animals' blood!

TONGUE-EATING ISOPOD AND FISH

These crustaceans are more like unwanted houseguests than friends. They attach themselves to a fish's tongue and make themselves at home in its mouth.

APHIDS AND ANTS

Aphids secrete a sugary liquid called honeydew. Some ants find it so tasty, they farm aphids for food! In return for "milking" their honeydew, the ants keep the aphids safe, cozy, and well-fed.

MANED WOLF

If you're ever wandering the savannas and wetlands of central South America, you may catch a whiff of this canine long before you spot it! A maned wolf marks its territory with stinky pee that smells a lot like skunk spray. *Pee-ew!* It's not related to skunks, though. And, despite its name, this animal is also not a wolf. It looks like a fox, with its longish legs and reddish fur, but it's not a fox, either. The maned wolf is truly an oddball, classified in a group all its own. It has another distinction, too—standing around three feet (90 cm) tall at the shoulder, the maned wolf is larger than any other doglike animal on the entire continent of South America.

Maned wolves will startle their prey by tapping a paw on the ground.

As much as 50 percent of the maned wolf's diet is made up of plant parts, including a tomato-shaped berry called lobeira.

LOWLAND STREAKED TENREC

The lowland streaked tenrec looks kind of like it's been electrocuted! Its coat is a mixture of fur and quills, giving it a prickly look. The quills make the tenrec a tough and painful meal! And they might be used for communication, too. The tenrec can shake and rattle the ends of some quills, making a squeaking sound. Scientists think tenrecs use this sound to find each other and stay together—especially at night.

VIOLET SEA SNAIL

This slimy snail goes with the flow. The violet sea snail spends its entire life drifting across the ocean surface hanging from a raft made of its own mucus. These mollusks cannot swim and do not move themselves in any way. Their course is always set by the wind. It cannot see and only eats the smaller animals that happen to drift by. You may think it's, uh, *snot* such a great way to live, but it seems to suit the snail just fine.

ACORN WOODPECKER

This bird has a bit of a hoarding habit. Groups of acorn woodpeckers work together to drill thousands of holes into a tree. The birds stuff an acorn into each hole, storing the nuts to be eaten over the long, cold winter. These birds have a unique community. Sometimes many generations will use the same tree to store their acorns. To make sure no one makes off with any acorns, members of the flock take turns guarding the tree.

No Way!!

CHEETAH

The world's fastest land animal can be hard to spot! Cheetahs can go from a standstill to about 60 miles an hour (100 km/h) in just a few seconds. How? These cats have a special spine. It flexes upward, bending in a rainbow-like arc as the animal leaps. This action gives the cheetah jumping power and allows it to cover lots of ground in a single bound. Scientists are fascinated by the cheetah's superstretchy spine and have used it as inspiration to build faster robots.

WALRUS

With their signature mustaches, large tusks, and roly-poly bodies, walruses have a truly distinctive look. If you watched them move around on a beach, you might think of them as clumsy creatures. Think again: Once these marine mammals hit the water, they go from weird to *wow!* Walruses are graceful swimmers, capable of reaching speeds faster than a person riding a bike. Most walruses dive underwater for five minutes at a time, but some can stay below the surface for up to 10 minutes, using their highly sensitive whiskers to feel for clams on the ocean floor. While underwater, the walrus's heartbeat slows. This helps the animal save energy so it can withstand the cold water temperatures of the Arctic Ocean.

Walruses drive their tusks into the ice and use them to pull themselves out of the water.

Walruses regularly gather by the hundreds on land. The largest group recorded topped 35,000 animals.

CRESTED PORCUPINE

Anything that manages to rattle this critter better back off. When a crested porcupine is threatened, its one-foot (30-cm)-long quills stand up, warning the predator. If the other animal doesn't take the hint, the porcupine shakes the quills on its tail, making a rattling noise that sounds like it's coming from a snake. It also stomps its feet and clicks its teeth. And if this still isn't enough, the porcupine runs backward at the attacker, quills first. But not everything about the porcupine is prickly. These rad rodents enjoy sunbathing on warm days.

LAMPREY

A lamprey's mouth can be a spooky sight. These fish have no jaws, but they make up for that with rows and rows of thorny teeth, arranged in circles. Lampreys swim up to other fish, bite into them, and hang on using their creepy chompers. The lamprey then feeds by sucking the blood and other liquids from inside its prey. Lampreys have monstrous appetites—a single animal can drain about 40 pounds (18 kg) of fish every year!

BLUE-FOOTED BOOBY

Look at those fancy feet! The vibrant color of the male blue-footed booby's webbed toes and legs is a source of pride for its owner. Males show off their blue tootsies in a mating ritual that includes a high-stepping movement. A female impressed with a male's high kick and brilliant blues will choose him for her mate. Later, when her eggs hatch, those funny feet will serve another purpose—they act like a blue baby blanket, covering the chicks to keep them warm.

BiZarre!

BUSH VIPER

This is one rough reptile. Though many snakes' scales look as smooth as silk, a bush viper's scales are pointed, jagged, and ridged. Scientists aren't completely sure why vipers have scales like this, called keeled scales, but think their spiky appearance may keep light from reflecting off the viper's skin, helping it stay unnoticed. But fancy scales aren't the only way these reptiles blend in. Bush vipers will hang upside down from a tree limb and stay still as they wait for prey. If a rodent, amphibian, or smaller reptile approaches, it's dinnertime!

WEIRDLY DECEPTIVE: HARMLESS OR DANGEROUS?

Sometimes you can tell if an animal is dangerous by its appearance. Sometimes, though, what you see isn't what you get.

SLOW LORIS

A loris's big eyes and slow movements make it seem harmless. But these animals have venom-producing glands, which they lick when under attack, and then bite their foe, injecting the toxin.

SCORPIONFLY

This insect's long pointed tail looks like it could deliver a wicked sting. The only thing it does, though, is help the male insect grab on to a female during mating.

HICKORY HORNED DEVIL

With its bright colors and spiky horns, this caterpillar looks downright devilish. But unlike many colorful insects, it's actually not poisonous.

61

GIRAFFE

Say *ahhhhh!* No, this animal wasn't licking a grape ice pop. A giraffe's tongue is naturally blackish purple in hue. This unusual color is thought to keep the giraffe's tongue from getting sunburned when it reaches above the protective shade of trees. Its tongue is also prehensile. Like a monkey's tail, the tongue is very mobile and can be used to grab on to and pull off the leaves that make up most of the animal's diet.

PUFFERFISH

You may know about pufferfish's unique ability to protect themselves by inflating their bodies into a prickly balloon. But did you know pufferfish have made their mating ritual into a fine art? Some species of male pufferfish create spectacular designs to attract females. They make a circle in the sand by dragging their bellies along the ocean floor. The artistic display continues as the fish swim in lines from the outside to the inside of the circle. Female fish choose the pattern they like the best and lay their eggs in its center.

pufferfish sand sculpture

IBERIAN RIBBED NEWT

Talk about side-splitting. When the Iberian newt is caught by a predator, it forces its ribs through its body. The sharp ends of the newt's bones pierce the predator's mouth, cutting it open. At the same time, the skin of the amphibian releases poison, which enters the fresh cut. All of this is no joke to the predator, which usually drops the newt, giving it time to get away. Fortunately for the newt, its skin heals quickly, so it doesn't seem to be too affected by literally poking its ribs out.

MEERKAT

School is now in session! Life on the African savanna is full of danger. Because a meerkat's diet includes venomous scorpions, every meal can be a matter of life or death. Adults teach their young pups how to hunt a stinging snack with some very realistic props. First, they bring a dead scorpion to the meerkat pup. Next, the adult will remove the stinger from a live scorpion and present it to the young meerkat. Once the pup has learned how to kill a stinger-less foe, it's ready to tackle the real, venomous thing. Class dismissed!

Glass frogs are about the size of a U.S. quarter.

GLASS FROG

Clearly, this amphibian has guts! The skin on the glass frog's belly and chest is see-through. Among the things you can see inside a glass frog's belly are its digestive system and its beating heart. And these frogs are really gutsy. They may be small, but they're fierce. The male frog, for example, dutifully watches over its mate's eggs. If an intruder, such as a wasp, comes too close, the frog will drive it off by delivering karate-like kicks with its hind legs. The female frogs lay their eggs on a leaf that overhangs a pond, lake, or stream. When the tadpoles hatch, they drop off the leaf right into the water below!

Glass frogs don't make the classic croaking sound many other frogs do—they make a high-pitched whistling sound.

ANGLERFISH

The deep sea is a weird world. In its inky black waters, where sunlight doesn't reach, you'll find the glowing lights of bioluminescent animals—or animals that can create their own light. One of these is the anglerfish. The female anglerfish has a spine curving above its face that is topped with a fleshy piece of glowing skin. Any animal that is hungry or curious enough to investigate will likely fall for the anglerfish's bait—hook, line, and sinker—and end up as a meal.

HUMMINGBIRD HAWKMOTH

Oh look, it's a hummingbird! Oh wait, no it's not. Hummingbird hawkmoths look and act like the birds for which they're named. Their wings flap at blinding speeds of up to 80 times a second, making a humming sound. Also, like hummingbirds, these insects are up and active during the day. One of the few moths commonly seen in the light, you can catch them darting from flower to flower, sipping nectar from their long proboscis mouthparts.

So Strange!

AFRICAN PANGOLIN

If you think this animal resembles an overgrown, walking pine cone, you're not alone. Pangolins are covered from head to tail with scales—the only mammal in the world to have them. They also have thin, sticky tongues that scoop up ants and termites. Poking out 16 inches (40 cm), the pangolin's extremely long tongue isn't attached to the back of its mouth. Instead, it's fixed to the inside of the animal's chest, where it's stored in a special sac until mealtime!

Rock and roll, baby! The palm cockatoo would be right at home on stage. Male cockatoos use sticks and seed-pods as drumsticks. They hold them in their beaks, then hit them against objects like tree branches in a steady beat. Why? To impress the ladies! And like any good musician, every male cockatoo marches to his own beat, too. Scientists have found that each male bird's rhythm is unique, which helps females tell potential mates apart.

71

WEIRD ANIMAL MYTHS

Earth's animals are plenty weird, but humans can make them even weirder by misunderstanding something they see. Here are a few myths you might have heard.

OPOSSUMS HANG BY THEIR TAILS!

Opossums are strong tree climbers, so it makes sense they'd hang on to branches by their tails, right? Wrong. These critters can only hang on for a few seconds as babies—adults are just too heavy!

TOADS CAN GIVE YOU WARTS!

False! Warts are contagious, but they aren't spread from animals to humans. Still, it's best to avoid poking at these critters. The bumps behind a toad's ears are known to irritate human skin! (And unwanted pokes are sure to irritate the toad, as well.)

SCARED OSTRICHES BURY THEIR HEADS IN SAND!

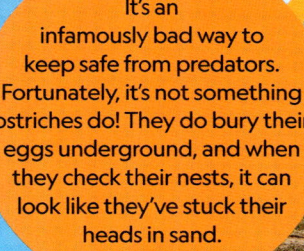

It's an infamously bad way to keep safe from predators. Fortunately, it's not something ostriches do! They do bury their eggs underground, and when they check their nests, it can look like they've stuck their heads in sand.

73

Naked mole rats often eat their own poop.

NAKED MOLE RAT

Naked mole rats can live for 30 years or even more, making them the rodent with the longest lifespan.

Mammals—they're fuzzy, furry, fluffy ... and naked?!? Mammals are associated with hair or fur, but not the naked mole rat! They lack the hairy coats that most mammals sport. But naked mole rats are not completely bald. They have little bundles of hair between their toes that help them push soil around underground, and whiskers on their faces to feel their surroundings. Another trait that sets them apart from many mammals? While most mammals live in groups, mole rats live in colonies—like bees! Each colony has a single female mole rat queen. The rest take on different roles, such as housekeepers that dig and repair tunnels, and soldiers that guard the nest and tell predators like snakes to buzz off.

SEA BUNNY

This bunny doesn't hop along—it oozes by! Although the sea bunny has a fuzzy, cuddly appearance, it's actually a sort of slimy mollusk. And those ears? They're more like antennae and are used to pick up smells from the sea slug's surroundings. The sea bunny's diet includes poisonous sponges. The slug's body becomes poisonous as it absorbs the toxins from its meals. So, as cute as this animal may appear to be, it's some-*bunny* that should be left alone!

HONDURAN WHITE BAT

As far as bats go, these really break the mold. Besides having bright white fur, Honduran white bats don't typically live in caves or attics. Instead, these mammals camp out in leaves. A group of bats choose a large leaf, then chew down its center to make the leaf fold into a cozy tent. As a bonus, when sunlight shines through the leaf, it makes the bats' fur look green, helping them hide from lurking predators like snakes.

WhAt?!?

SCORPIONFISH

This fish is quite an eye-catcher! The eyespots on the scorpionfish's fins might make it appear to be especially alert, but the spots actually serve another purpose. Predators looking to attack the head of this creature might go for its fins instead of its head. This wrong move gives the scorpionfish time to scoot away. By the way, you wouldn't want to step on this bottom dweller. Its back fin is full of prickly spines that deliver a super-painful—and toxic—sting.

You all right there? "Playing possum" means pretending to be dead. And these mammals play dead better than most. Opossums can pose with their tongues hanging out for hours at a time. But that's not all! They have some gross tricks up their sleeves, including foaming at the mouth and oozing smelly green slime from their rear ends. These tricks may signal to predators that the opossum is sick and not a great choice for a meal.

These long-armed animals aren't great at walking around, but they're surprisingly strong swimmers.

SLOTH

Don't rush me! Sloths are famously slow and almost never touch the ground, except for potty breaks. But when they go, they make it count. Some sloths poop out a third of their body weight in one go and only poop once a week. Sloth potty trips are also part of an incredible life cycle. A sloth's fur is home to green algae (which the sloth eats) and moths. The moths lay eggs in the sloth's poop. Soon the eggs hatch, and the larvae grow into adults, which fly back to the sloth's fur, bringing some of the materials in the poop with them. These materials fertilize the algae. Throughout this amazing cycle, the sloth's fur is like a farm for healthy, nutritious algae!

Sloths can be either two- or three-toed, with either two or three long claws on each limb.

81

COUGAR

What's this big cat's name? Well, it depends on who you ask! In different parts of the world, it's called a puma, mountain lion, panther, catamount, or cougar—but it's all the same animal. Along with its many names, this big cat comes with some unusual traits, too. All cougar cubs are born with bright blue eyes, for example, and their call sounds very much like a person screaming. They're also expert jumpers and can leap forward 40 feet (12 m) in a single bound. That's about the length of two pickup trucks!

Hey, nice hat! A Brazilian treehopper's helicopter-like decoration is certainly eye-catching. The strange-looking structure grows from the back of the insect's middle section, called its thorax. But, oddly, no one seems to know exactly what it does. Some scientists think that these bizarre balls might make predators think the treehopper has a disease-causing fungus growing on it. That's a pretty impressive defense for an insect that's only about the size of a pea!

BiZarre!

STONEFISH

Yes, there is an animal in this picture! Reef stonefish are almost invisible when lying next to a coral reef. They can stay still for days at a time, moving only when a fish swims by their enormous mouths, then—*gulp!* Because they don't move around very much, stonefish might seem like good targets for animals clever enough to see through their camouflage. That's the opposite of true. A row of spines down this animal's back contains some really powerful venom—enough to make any animal that's stung feel like it's been hit with a hammer.

SCREAMING HAIRY ARMADILLO

This little armadillo is named for the screeching sound it makes when it's in danger, but that's not its only strange trait. For one, these desert dwellers hunt for prey by pushing their heads into the sandy soil, before turning their bodies in a circle to make a hole. This method means that any meal comes with a hearty helping of sand. Not that these armadillos are particularly picky—they also feed on dead animals and garbage.

Platypuses don't have teeth, so they scoop gravel into their mouths and use it to grind up their food.

There are about 40,000 receptors on the bill of one platypus.

DUCK-BILLED PLATYPUS

No roundup of outrageous animals would be complete without the duck-billed platypus. The first scientists who laid eyes on one thought it was a fake. They found the creature so odd that they assumed someone was playing a joke on them! The platypus has so much weirdness to love beyond its unique bill. It lays eggs, one of only a few mammals to do so. Also, females don't have nipples—they "sweat" milk for their young to lap up. Males have venomous spurs (small, sharp barbs) above their back feet that can deliver a painful sting if attacked. And about that snout: Although it is shaped like a beak, it is soft and covered with thousands of sensors that can detect tiny electrical impulses given off by insects and other prey.

DRACO LIZARD

Draco lizards would look at home in a book of wizards and magical creatures. Although they don't truly fly, these reptiles are excellent gliders. Some of their ribs are extra long and connected by skin. These winglike flaps are usually folded against the lizard's sides. When the lizard leaps, it unfolds the flaps and glides to a tree, using its long tail to steer. Draco lizards regularly glide between objects about 30 feet (9 m) apart, about the length of a school bus. But they can soar up to 160 feet (50 m), the length of an Olympic swimming pool!

At first glance, this little bird might look just like any other songbird. Look a little closer, though, and you'll notice its oddly twisted beak! A crossbill beak's top and bottom overlap in a curved X-shape. The ends of the bird's bill are just the right shape for squeezing between the overlapping scales of pine cones to pull out seeds. Crossbill chicks are born with normal triangle-shaped beaks. As they grow and feed, the tips lengthen rapidly and morph into their peculiar crossed shape.

WhAt?!?

HAMMERHEAD SHARK

Good underwater vision? Nailed it! A hammerhead shark's long, flat head looks a little strange, but the position of its eyes on either side lets the shark see more of the world around it than other shark species. And that's not the only perk of this creature's bizarre noggin. Like all sharks, hammerheads have structures on their skin that can sense electricity emitted from potential prey. Hammerheads have more of these special structures than other sharks, so they can sense more of these signals at once.

DUNG BEETLE

Dung beetle larvae hatch inside a dry ball of animal poop and then eat it. Adult beetles slurp up nutrient-rich liquids often found in waste left behind by herbivores, or plant eaters.

BEAUTIFUL WOOD NYMPH

Resembling poop is a lifesaving defense for this wood nymph. While resting, their wings are folded overhead and their legs outstretched, making them look like bird droppings, which is enough to keep predators away.

93

TARSIER

What are you looking at? If a person's eyes took up as much space in their head as a tarsier's, they'd be about the size of grapefruits. In fact, this creature's eyes are so large that it can't move them. Luckily, tarsiers have superflexible necks and can turn their heads 180 degrees in each direction to look around. That's nearly as good as having eyes on the back of their heads! Tarsiers' staring, saucer eyes help them see better in very dim light. The nocturnal critters also have suction-cup-like toes—a feature that helps them make a safe and steady landing as they leap from tree to tree in their Southeast Asian forest habitat.

The tarsier is the only primate that eats only meat, including insects and small reptiles.

The tarsier is only about five inches (13 cm) long, not including its tail.

95

WhoA!

HUMPBACK WHALE

Have you ever blown a bubble or belted out your favorite song? If so, you may be more similar to a humpback whale than you think! Groups of humpbacks swim circles around schools of small fish, blowing bubbles as they go. The bubbles herd the fish into one spot, so it's easier for the humpbacks to scoop them up. Male humpback whales are also famous for their singing. They will make a long, complex series of moans, whistles, and howls that some scientists think can be heard thousands of miles away—across entire oceans!

PINK KATYDID

If blending into the background is key to survival, it seems that this insect might not have gotten the message. Most katydids are brown or green, and easily hide in trees, but not the pink katydid. It stands out in the leaves, making it a fairly easy target for birds and other predators. Why are some katydids pink? They inherit the trait from their parents. The color results from a mutation, or unusual change, in the critter's genes, which scientists don't fully understand. However, only about one in every 500 katydids has this distinct pink coloring, which means they're quite rare.

This antelope looks pretty odd, and it has some strange habits to match. Gerenuk roam the dry savannas of eastern Africa. They don't eat grass but munch on other plants, flowers, and fruit. This diet is pretty juicy and provides plenty of moisture, so they don't have to drink water. The gerenuk marks its turf with a substance made by certain glands on its body. When it rubs against branches or trunks, the gerenuk's scent is left behind, letting other animals know that the area is spoken for. The animal also has glands behind its eyes that make a smelly substance that looks like tar.

THORNY DEVIL

A thorny devil's scales have a lot of jobs to do. In addition to making this small lizard look super spiky, the scales help the animal stay hydrated in the Australian desert. The scales act like miniature sponges, soaking up water from damp sand and morning dew. And, if you think that this animal looks like it has two heads, you're not seeing double. The thorny devil has a soft knob of tissue on the back of its neck. If attacked, it tucks its real head between its front legs, while the false one sticks up, to try to fool the predator.

AXOLOTL

In the past, human explorers searched for a fountain of youth to no avail. But the axolotl seems to have found it. Most salamanders go through metamorphosis, changing from a young form into an adult one. This means losing gills that help them breathe in water and gaining lungs that help them survive on land. Not this amphibian. They keep their youthful gills but grow lungs to breathe air, too. An axolotl also has the superhero-like ability to regrow parts of its body, including legs, parts of its heart, and even some of its brain. It happens quickly, too! This amazing animal can flawlessly regrow a leg five times within the span of a few weeks.

In the wild, these critters can only be found in a few lake systems around Mexico City.

This salamander's name means "water dog" in an Indigenous Aztec language.

PROBOSCIS MONKEY

The male proboscis monkey's nifty nose looks, well, a little odd to most people. But in the jungles of Borneo, these noses are all the rage, at least among proboscis monkeys. The oversize nose acts like an amplifier, making the monkey's call louder. This impresses female proboscis monkeys, which pair off with males that have loud calls and big sniffers. The monkeys are also known for being great swimmers but terrible divers. They often enter the water by leaping from a tree, landing flat in a belly flop.

CUYABA DWARF FROG

This frog's bottom is bizarre—no ifs, ands, or *butts!* Cuyaba frogs attempt to scare off predators by puffing up their rear ends. The large dark spots on the amphibian's backside look a lot like the eyes of a larger animal and may cause predators to think twice about attempting to attack. If the frog's predators, which include other South American animals like snakes and coatis, ignore the first warning, the animal has a backup, backside plan. The skin under those dark spots releases toxins that can harm the predator.

LYREBIRD

What's that sound? Lyres are small U-shaped harps, used to play sweet, soft sounds. But lyrebirds, which got their name because their tails look like the musical instrument, are one of the noisiest animals in the Australian forest. They're also copycats—mostly imitating other birdcalls. A single bird can learn to mimic the calls of more than 20 other birds. And the lyrebird's skills don't stop there. People have heard them doing spot-on impressions of objects such as car alarms and chain saws.

Lions are so iconic that it might seem strange to call them "weird." But in the feline world, they are quite peculiar. Living in complex groups called prides, lions are the most social of all cats. Each pride is made up of female adults and cubs, sometimes reaching as many as 40 lions. Each pride also includes one or more males that defend the pride while the females take care of the cubs and most of the hunting.

SAIGA ANTELOPE

Saigas can run 50 miles an hour (80 km/h) when they're fleeing danger. That's about the speed a car travels on a local highway!

With hot, dry summers and chilly winters, life in the northern grasslands of Asia can be harsh. Enter the saiga's strange schnoz. This mammal's big nose filters out dust kicked up by the hooves of other antelope. Its large size helps the animal stay cool by releasing heat. In winter, the large nostrils heat up the air as the saiga breathes in, keeping the antelope warm. And that's not all there is to know about this nifty nose. Scientists think it might be related to the animal's great sense of smell and may even help make its "roars" louder during mating season—both of which help the antelope thrive in its environment. The nose knows!

These antelope can travel up to 75 miles (120 km) every day during their yearly migrations.

VERREAUX'S SIFAKA

Sifakas are part of the leaping lemur family, so as you'd expect, they are amazing jumpers. They grab on to tree trunks or stems and can spring as far as 33 feet (10 m) from tree to tree. Even when on the ground, they jump along on two feet rather than moving on all fours. Leaping lemurs!

Parrotfish ... peculiar? Positively! The fish uses its bony parrotlike "beak" to grab and break off chunks of coral from a reef. Then, plates inside the parrotfish's mouth grind up the hard outer material so it can eat the softer part of the animal. We have this process to thank for some of the sand we enjoy at beaches around the world. When they eat coral, parrotfish poop out sand! So, the next time you're relaxing at a white sand beach, you might be enjoying the hard work—and poop—of parrotfish.

DUMBO OCTOPUS

Listen up! The big flaps of skin on the Dumbo octopus's head aren't ears. They're fins! The mollusk uses its fins to push itself through the water and steers with its arms as it coasts just above the seafloor looking for food. No other octopus lives deeper in the ocean than the Dumbo octopus—it's found in depths of up to 13,000 feet (4,000 m) below the surface. That's more than twice as deep as the Grand Canyon! It's so deep that the Dumbo octopus almost never comes across a predator.

So Strange!

ASSASSIN BUG

This bug comes with a built-in straw, but it's not used to slurp up milkshakes. An assassin bug stabs smaller insects with its proboscis, a long, hollow mouthpart. The assassin bug then injects a toxic liquid into its prey before sucking it dry. Gross? Yes. But some kinds of assassin bugs take it one step further. These critters make a gluey substance and use it to stick the empty shells of their victims to their backs. Scientists aren't completely sure why, but they think the bugs use this as a disguise so predators don't recognize them.

111

NARWHAL

Weird ... or magical? It's clear why narwhals are sometimes called "unicorns of the sea." But what looks like a horn from a storybook isn't truly a horn. It's a long, overgrown tooth called a tusk. And like your own teeth, it's sensitive. No one is exactly sure what the purpose of the tusk is (other than looking super cool). Scientists think it may be used to pick up the movement of fish swimming by. They have also observed narwhals smacking their horns into schools of fish. Fish that collide with the horn are temporarily stunned and make an easy meal for the narwhal. Males might use them for dueling. Females, for the most part, don't grow tusks.

A narwhal's spiral-shaped tusk can grow to be about 10 feet (3 m) long.

Some narwhals can even grow two tusks.

WOMBAT

Most scientists estimate that there are at least eight million kinds of animals on Earth. Only one has poop shaped like a cube, though—the wombat. These Australian nocturnal mammals poop about a hundred times a night. The poops look like the perfect building blocks—and that's just what wombats use them for. They make stacks of doo-doo on the edges of their territory, warning other wombats to stay away. Any wombat that ignores the poop pile risks a fearsome attack.

wombat poop

MARINE IGUANA

Marine iguanas are the only reptiles that live near the ocean. The plants they eat are soaked in salt, which would be a problem for most lizards. But not marine iguanas! These iguanas have special glands where the extra salt in their bodies is collected. They then blow the salt out of their noses with great force, something that looks and sounds to people like an especially wet sneeze. And because the iguanas don't use tissues, the salt tends to build up in a white crust on their faces. Bless you!

No Way!!

BURROWING OWL

Most owls hoot from their nests in hollow trees. Not burrowing owls—they hiss like rattlesnakes from their cozy holes in the ground! It's a great way to scare off would-be predators. They're also messy housekeepers, collecting other animals' poop and scattering it around the entrance of their burrows. It sounds yucky, but it acts as a home meal delivery service. The poop attracts insects, which the owls then catch and eat. Some animals decorate the outside of their holes with bits of trash like bottle caps, too. Scientists think this shows other owls that the burrow is occupied.

RED LAND CRAB

This is quite the animal crossing! Cuba is an island country located where the Gulf of Mexico and Caribbean Sea meet. Every year, millions of migrating land crabs make their way from the forests that grow in the middle of Cuba to the island's shore. The crabs can travel up to six miles (10 km) over several days. The journey is tough and dangerous. Many of the crabs don't make it, ending up as food for seabirds or unable to make it across roadways. Recently, human-made tunnels and road bridges have created safer routes for them. Crabs that do reach their destination spawn in the sea.

117

ELEPHANT SEAL

Elephant seal pups are an incredible 75 pounds (34 kg) at birth!

Male elephant seals are truly enormous. Some can weigh up to 8,800 pounds (4,000 kg)—that's as much as a small truck. Only males have the long noses that give the animal its name. They use these significant schnozzes during mating season. By inflating their noses, the male seals make sounds that can be heard from miles away. The loudest, most dominant males can have more than 100 mates at one time. Both male and female elephant seals possess some other exceptional traits: They're super swimmers, reaching speeds that top 10 miles an hour (16 km/h). They can even hold their breath underwater for more than two hours during a dive! It's no wonder that they only come onto land to shed their fur and to have babies.

These mammals will dive 1,000 to 2,000 feet (300 to 600 m) while looking for fish, squid, and small sharks to eat.

GOLDEN TORTOISE BEETLE

For a unique way to change color, this insect wins the gold. The shells of golden tortoise beetles are made of many layers. Fluid between the layers reflects light so the beetle appears gold. When threatened, though, some of that fluid dries up. This makes the animal change from gold to a dull red. The reddish color makes the tortoise beetle look as if it might be poisonous to eat, so predators are more likely to leave it alone.

What can get to be 60 feet (18 m) long, glow in the dark, and look like it should be waving outside a store that's having a sale? The giant pyrosome! What looks like one long tubelike animal is really a colony of thousands of much smaller animals. Each individual animal draws in water from the outside of the tube, filters out tiny bits of food, then pushes the water further inside the tube. The water squirts out the back of the pyrosome, propelling the entire colony forward.

LEAFY SEADRAGON

Where you might see an animal, a predator sees seaweed! Leafy seadragons are small seahorses that live in shallow waters off the coasts of Australia and New Zealand. These animals definitely go with the flow. Their fancy fins aren't good for swimming, so they mostly drift along with the ocean currents, making them look even more like seaweed. Weird, but it works—most predators swim right by without a second glance.

NORTHERN SHRIKE

This bird takes the idea of "waste not, want not" to grim extremes. Northern shrikes are always on the hunt: If there's a chance to snag prey, like a smaller bird or a mouse, they take it. But what if they catch something on a full stomach? No problem, because the shrike doesn't mind leftovers. It'll just find a sharp object like a spiky twig, stick the food on it, and leave it there until snack time. Storing food for a day or two also dries it out, making it easier for the shrike to eat.

WEIRD EARS

Let's hear it for ears! These body parts help animals hear, and sometimes they help animals survive in other ways, too!

FENNEC FOX

Small fox, big ears. The fennec's big ears help direct the intense desert heat away from the fox's body.

GOBI JERBOA

When compared to its entire body size, this tiny rodent's ears are the largest in the animal kingdom.

KUDU

This spiral-horned antelope has ears like satellite dishes. They rotate them in the direction of even the faintest noise to detect hidden predators.

CARACAL

Here's a tip—the tufts of fur that crown a caracal's ears aren't just for looks. They direct sound into the cat's large ears, helping it find prey.

COCKATOO SQUID

Oh, what big eyes you have! The cockatoo squid lives about 1,600 feet (500 m) deep in the ocean, with some probably dwelling even deeper. Not much light reaches that far, so the squid's huge peepers help pick up as much light as possible. Also, this squid's body is see-through, which is why it's also known as a glass squid. The squid's body is filled with ammonia, a clear chemical that many animals get rid of in their pee as waste. But ammonia, which is lighter than seawater, helps the squid stay afloat. Odd ... and resourceful!

Have a heart! The red pouch over the frigatebird's chest resembles a cartoon heart. During breeding season, the bird inflates the pouch with air. Sometimes the pouch becomes as big as a person's head! When it comes to competing with other birds for food, though, the frigatebird doesn't share the love. This animal gets the nickname "man-o-war bird" because it steals food from other birds midflight. A frigatebird will pester another seabird that has recently had a successful hunt until it either drops or throws up its meal, which the frigatebird grabs before it hits the ground.

MARGAY

Many wild cats like to hang out in trees. The margay likes to literally hang *out of* trees. These small cats native to Central and South America have extremely flexible and strong ankles. Margay cats can dangle from a tree limb using just one foot. They can also rotate their ankles about 180 degrees. Backward-facing feet mean these cats can walk down the trunk of a tree headfirst!

COOL!

SHORT-HORNED LIZARD

The things animals do to defend themselves can be "bloody" weird. The short-horned lizard is fairly well protected by its spiky scales. When threatened, though, the reptile's body can inflate to nearly twice its size, which can scare off a predator. If this fails, the blood vessels near the reptile's eyes burst from pressure and squirt blood at the attacker. The jet of blood tastes and smells bad to common predators like coyotes and can hit a target about three feet (1 m) away!

129

PISTOL SHRIMP

Don't burst this shrimp's bubble! A pistol shrimp has a special claw that launches bubbles. That may sound cute, but it's serious business—the bubbles are released at speeds of over 60 miles an hour (100 km/h). This comes in handy when the shrimp is hunting—just the sound of these bubbles stuns small fish with enough force to knock them out. In fact, the popping sound of the bubbles is louder than the shot of an actual pistol and can be enough to interfere with human-made tech, like sonar and other underwater devices.

Everyone who thinks this animal is a little weird-looking, say "aye!" Found only in Madagascar, aye-ayes are primates, which means they are related to monkeys, apes, and people. According to local lore, spotting one is bad luck. Some people are so wary of the aye-aye that even speaking its name is a no-no. In truth, aye-ayes are harmless, even if they are more than a little odd. They're only active at night, and they have big, sensitive ears, eye-catching eyes, and long, spindly fingers. To hunt, aye-ayes tap a finger on a tree branch, then listen for the sounds of insects crawling underneath the bark.

Lion's mane jellyfish are bioluminescent, meaning they can create their own light and glow underwater.

LION'S MANE JELLYFISH

The sight of this animal is pretty hair-raising! The lion's mane jellyfish's hairy-looking tentacles are full of special stingers. The stingers release a powerful toxin, stunning prey such as small fish, shrimp, and other jellyfish. Even without its tentacles, this creature is monstrous in size. Its body is usually about 16 inches (40 cm) in length but can be as long as six feet (1.8 m)—sometimes even longer. Its fabulous "mane" can grow as long as 120 feet (37 m). That's close to the size of a blue whale! Despite its size and ferocious name, the lion's mane is sometimes prey for other animals. Sea turtles, for example, aren't bothered by the tentacles' sting and won't hesitate to make a meal out of the big jellyfish.

HONEY ANT

When honey ants, also known as honeypot ants, get hungry, there's no need to order takeout! In honey ant colonies, some of the insects act like pantries. Most of the ants collect a sweet food called honeydew and store it in their bellies, which swell up as the insects get full. The full ants latch on to the ceiling of a tunnel and hang out until the colony needs food. Then, they come down and spit up the honeydew for whoever's hungry. How sweet!

WEIRd!

This animal's hood doesn't help keep it warm. The male hooded seal has a stretchy flap of skin on the front of its head. When the animal wants to attract a mate, it seals one of its nostrils closed. It blows air out of the nostril and into the skin. The flap fills with air like a balloon—and looks like one, too. Some of these flaps can get to be about the size of a football! This also acts as a warning sign of aggression to tell other males to back off.

GUANACO

Guanacos laugh in the face of danger! Literally. As these animals graze, a few individuals take turns standing on higher ground, acting as guards for the larger group. If they spot danger, they let out a call that sounds a lot like laughter. This warns the other guanacos that it's time to run. A guanaco's other response to danger is to spit. They throw up the contents of their stomachs and spit them at the attacker's face. (We're not joking!)

BOBBIT WORM

Beware what lies beneath. A bobbit worm is an alien-looking critter that buries most of its body in the sand of the ocean floor. The only thing visible? Its five antennae, which poke above the ground, reaching for anything that swims by. When a possible meal is in range, the worm launches its head out of the sand and clamps its jaws on the prey. Sometimes the force is enough to cut its meal into two pieces.

137

COLOR POP!

Many animals have hues that help them blend into the background. But some animals really stand out! Here are a few that aren't afraid to show their truly colorful selves.

SPLENDID FAIRY-WREN

For most of the year, the fairy-wren's feathers are a rather dull shade of brown. But during breeding season, male fairy-wrens get the blues. Their feathers turn all shades of brilliant blue!

PINK HAIRY SQUAT LOBSTER

This animal is pretty in pink. Hairy squat lobsters vary in hue from pink to purple to orange to red and brown. They have powerful pincers that they use to grab prey.

MANDRILL

The mandrill's colorful face makes it look like it's wearing clown makeup. But these monkeys wouldn't make good performers—they are actually very shy.

OKAPI

Now you see them ... now you don't. Okapis are notoriously shy and hard to spot. Unlike their more famous relatives, giraffes, okapis are built for life in the rainforest. Their striped legs and backsides help them blend into leafy shadows. Their fur is also like a built-in raincoat—it's thick and oily, so the water slides right off! They spend their days munching on rainforest leaves, with the help of their superlong tongues. But when they need more nutrition than leaves can provide, okapis have been known to snack on clay and even bat poop! Their exceptionally long tongues are also perfect for washing their ears and eyes.

Solitary okapis mark their territory with each step, as glands in their feet leave a scent behind.

Okapis make sounds that are so low in pitch that humans cannot hear them.

141

BEE HUMMINGBIRD

Want to know the buzz about this bird? It's so unusually tiny that it's often mistaken for a bee or other insect. It measures only about two inches (5 cm) long and weighs less than 0.1 ounce (3 g). Females build nests that are only about one inch (2.5 cm) wide and lay eggs that are about the size of coffee beans. If you want to see the oddly tiny bird for yourself, though, you'll have to go to Cuba. It's the only place the bee hummingbird is found.

No Way!!

BOTTLENOSE DOLPHIN

Bet you didn't know this marine mammal has skills worthy of a super spy! Dolphins communicate with one another in many atypical ways, including tail slaps, jaw claps, and even by blowing bubbles. A bubble ring might signal it's time to play, for example, while a bubble stream might mean danger. Sometimes dolphins make these bubbles silently. Other times, the shapes are accompanied by a whistle that can be heard underwater. Scientists don't fully understand this behavior, so they'll continue to study the dolphins in hopes of cracking the code.

MATAMATA

The matamata's long, tubelike snout looks a little like a snorkel and acts like one, too! Like all reptiles, the matamata has to breathe air, but it can do that by poking the very tip of its nose above the surface of the water. This master of disguise also has an unusual head, neck, and shell. They are brownish and ridged, making the turtle look like a fallen log as it sits motionless in the murky water. When it sees likely prey, it opens its jaws, sucking water—and hopefully a meal—into its wide mouth.

Is there a weirder animal than this one? Only when pigs fly! This sea cucumber, aka sea pig, absolutely hogs the limelight with all its odd traits. For starters, sea pigs breathe using their butts! As they scoot along on their 10 to 14 "legs," they draw water up into their bodies through their bottoms. A special body part removes the oxygen before pushing the water back out. Some sea pigs even use their rear ends as second mouths, as if the one surrounded by tentacles that collect food—which includes tiny pieces of dead animals and plants—wasn't enough! Predators think twice before pigging out on this animal. It's poisonous!

JAPANESE MACAQUE

Pink-face Japanese macaques know about self-care. These animals, also called snow monkeys, are found only in Japan. No other primate (except humans) lives farther north, so having a plan to deal with the cold is essential. During the winters, macaques lounge in natural hot springs. The spa treatment not only keeps the macaques warm but it helps them combat stress. Scientists know this because studies showed that macaques that take longer baths have fewer stress-related chemicals in their poop. The macaques don't just soak in water, either. They're ace swimmers and will do laps in the hot springs. Macaques can paddle about a third of a mile (0.5 km) without a break.

The water these monkeys bathe in is a toasty 100°F (38°C).

Macaques know how to have fun in the snow, too—youngsters have been spotted having snowball fights!

WEIRDLY EXPRESSIVE

You've probably heard that a picture is worth a thousand words. Well, that's true for animals, too!

GOAT

GREAT WHITE SHARK

DAMSELFLIES

CHIMPANZEE

149

MANDARINFISH

Beast or beauty? A tropical mandarinfish's bright colors are a warning to possible predators—don't touch! Mandarinfish don't have scales for protection like most fish do. Instead, their bodies are covered with small prickly spines, making the fish more than a little difficult to choke down. And if that's not unappetizing enough, the spines also make a mucuslike slime that smells horrible and is toxic to other animals. The mandarinfish's mating ritual, though, seems right out of a fairy tale. Females gather to watch males perform a sort of dance just after sunset. How romantic!

TAPIR

Is it a ... pig? A small hippo? None of the above! The tapir is a confusing-looking animal. It's related to rhinos and horses but has a flexible, mobile nose that it uses to grab on to leaves and stuff them into its mouth. The tapir is a solid swimmer and when it's just cooling off in the water, it will poke its snout above water like a snorkel. Another reason for the tapir to go in the water? It often poops there! This is probably a way of hiding the smelly stuff from jaguars and other predators, keeping the tapir's location a mystery.

Batfish are so named because of the flat, batlike shape of their bodies.

RED-LIPPED BATFISH

M *wah!* Red-lipped batfish look ready to pucker up. These fish, which only live near the Galápagos Islands in South America, have bright red smackers. Scientists aren't sure, but some think this colorful mouth helps the fish attract a mate. And what about those "legs" sticking out from its sides, you ask? Well, this creature also does something most fish don't—it "walks." The batfish can swim but often uses its fins like legs to walk on the ocean floor or to perch on a coral reef, waiting for a possible meal to swim by. With its awkward way of moving, the batfish is not the greatest hunter. But the adult fish has a fleshy knob called an illicium on its head that is especially effective at luring prey.

> Around the world, there are more than 50 species of batfish.

CLOWNFISH

Clownfish and anemones are pretty perfect roommates. The anemone looks like a flower, but it's not. It's an animal, one that defends itself with the stinging barbs inside its tentacles. Clownfish, though, don't feel the anemone's sting. The fish take shelter in the anemone's tentacles but aren't harmed because clownfish have a layer of slimy mucus that forms over their scales. The slimy stuff protects the fish, keeping them safe from the anemone's barbs. It's a win for the anemone, too: Clownfish help clean the anemone and chase away predators.

BLACK HERON

This bird has it made in the shade! A black heron lures possible meals right to its feet by offering prey a nice spot to cool down. It raises its wings up in the air, causing a shadow to form. To unsuspecting fish, it's an appealing place to rest—until the heron strikes with its open beak, scooping up its meal and swallowing it whole.

BOMBARDIER BEETLE

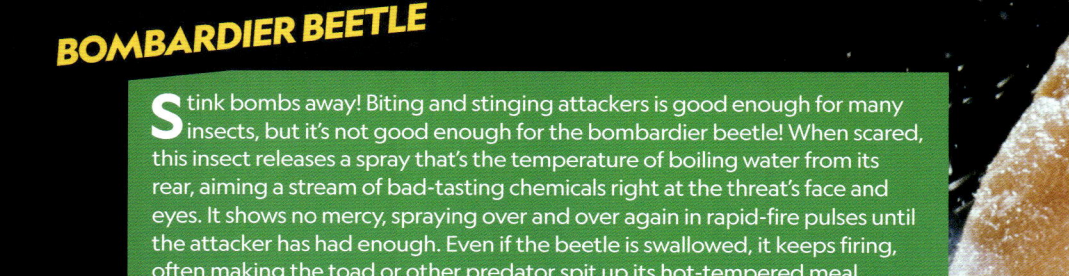

Stink bombs away! Biting and stinging attackers is good enough for many insects, but it's not good enough for the bombardier beetle! When scared, this insect releases a spray that's the temperature of boiling water from its rear, aiming a stream of bad-tasting chemicals right at the threat's face and eyes. It shows no mercy, spraying over and over again in rapid-fire pulses until the attacker has had enough. Even if the beetle is swallowed, it keeps firing, often making the toad or other predator spit up its hot-tempered meal.

WhAt?!?

CAMEL

This animal looks ready for its close-up. Camels aren't satisfied with two eyelids—they have three, and their eyelashes do way more than make them look glamorous. The two rows of long hairs help trap blowing sand, keeping it from getting in the camel's eyes. A pair of rather bushy eyebrows have the same job. Long nostrils act much like filters on a car do, preventing the camel from getting a nose full of sand and dust when it breathes in. Turns out a hairy situation can be a good thing ... if you're a camel!

157

These geckos live only in the southeast part of Madagascar.

SATANIC LEAF-TAILED GECKO

The satanic leaf-tailed gecko has more than one trick up its sleeve! First off, animals hoping to make a snack of this rainforest reptile are going to have a tough time finding it. Its head has scaly skin above its eyes that looks like tree bark, and its tail looks just like a rotted leaf. Also, when the gecko flattens its body against branches, its shadow shrinks. If these camouflaging tricks fail to work, the gecko has another strategy to avoid becoming a meal. It opens its wide red mouth and lets out a terrifying scream!

Satanic leaf-tailed geckos sleep hanging upside down by their tails!

159

PINK DRAGON MILLIPEDE

Imagine coming across this critter on a hike! The pink dragon millipede hangs out in cool caves and leafy forests in places like China and Thailand. It's only about the size of a grape, but the dragon part of its name shouldn't be taken lightly! Its pretty pink color is a warning sign that the millipede is not a great snack. Take heed, predators: This creepiest of crawlies can release cyanide, a deadly poison, to protect itself.

Solenodons have a reputation for their bad, bad tempers. Most of the time, they will ignore other animals, instead concentrating on sniffing along the forest floor for food. But if threatened, they go from relaxed to enraged almost instantly. Despite being small, and having an extremely poor sense of sight, the solenodon isn't a pushover. It is one of the world's few venomous mammals, and it will take on amphibians and reptiles larger than itself, disabling them with a single bite.

QUOKKA

Say cheese! The quokka looks as if it is always ready to get its picture taken thanks to its upturned mouth. The cute quokka isn't actually happy all the time. Its oversize front teeth and the arrangement of its face muscles make it look like it's grinning. Although it is best known for its smiley expression, the quokka's feeding habit is also unique. It regurgitates its food over and over to make sure it is getting all the nutrients from the water it drinks and the plants it eats. Yum!

WHOA!

ORANGE FIDDLER CRAB

There's no need for this crustacean to claw its way to the top. The male fiddler crab is known for its unevenly matched claws. It has one that is much bigger than the other. The crab mostly uses this larger claw to communicate with other crabs, waving it back and forth to attract mates or warn enemies to stay away. The orange fiddler crab's oversize claw is really big—it can make up as much as half of the animal's body weight! The crab itself only measures 1.6 inches (40 mm) across. It's all relative!

163

PEACOCK MANTIS SHRIMP

Mantis shrimps are about seven inches (18 cm) long.

Don't let the peacock mantis shrimp's bright colors and tiny size fool you into thinking it's harmless. It packs a punch that some scientists say is stronger and faster than any other animal's strike. The colorful mantis shrimp doesn't fight for sport, though. It uses its clobbering power to stun—and sometimes dismember—prey. If the shrimp comes within reach of a crab or small mollusk, it's likely to be lights out (and shell broken) for the other animal. Peacock mantis shrimps can throw a punch hard enough to shatter glass. Clocked at 50 miles an hour (80 km/h), the punch is fast enough to cause an audible pop and a flash of light!

These vivid crustaceans have very complex eyes that can see more colors than people can.

ANKOLE CATTLE

No need for these animals to toot their own horns—their headgear is obviously impressive! These cattle were originally from Africa, but they're now found all around the world. Their impressive horns are the biggest around, no matter how you measure them. They can grow to almost three feet (1 m) long, and their tips can be as much as six feet (2 m) apart. Fortunately for the cattle, the horns are hollow, so they aren't very heavy. Long blood vessels run along the inside of each horn, helping the animal stay cool.

ARMADILLO GIRDLED LIZARD

Have you ever heard the expression "put your foot in your mouth"? This lizard takes it further! When threatened, an armadillo girdled lizard will put its tail in its mouth and curl up into a ball. In this position, only its thorny scales are exposed to the danger. This lizard is named after the armadillo, which uses the same defense mechanism. The reptile can also detach its tail if it's grabbed by a predator, but it will only do that as a last resort.

167

CANADA LYNX

The Canada lynx's mismatched legs make it look kind of funny, but its hunting skills are definitely *snow* joke! The cat's hind limbs are longer than its front ones, an adaptation that helps the lynx jump over snowdrifts as it hunts in the pine forests of North America. A lynx's paws might look a little too big for its body, but they're also helpful for the hunt. They act like snowshoes, keeping the lynx from sinking into the snow. For these cats, there's really *snow* place like home!

So StranGe!

OLM

No eyes, no problem. Olms swim in water found in the dark caves of European countries such as Croatia and Slovenia. When this salamander is fairly young, its eyes stop developing, and over time, they're covered by skin. The amphibian relies on other super-strong senses—including the ability to detect electric fields—to make up for its lack of sight. Olms tend to eat very little—some eat only once every 10 years!—and they don't move around much. Scientists noted that one particular animal didn't move from one spot for seven years.

BIZARRE BEAKS

Bird beaks can be short and fat, long and thin, and all colors of the rainbow. Still, some birds have beaks that are just a little weirder than most others. Here are a few of them.

PUFFIN

Puffins' gray beaks bloom orange in spring! If puffins find fish to eat on a dive, they carry all they can at once—definitely a "beak-ful"!

SHOEBILL

This bird uses its impressive hooked bill, which resembles a wooden clog, to gobble up fish, lizards, and even small crocodiles.

SKIMMER

It looks strange, but the black skimmer's beak is perfect for its hunting style. Flying with just the lower part of its bill under water, it snaps up any fish it touches.

171

COATIMUNDI

It's not going out on a limb to say that a coatimundi (or coati, for short) is adapted for life in the treetops. Take, for example, its habit of walking down the trunk of a tree headfirst. The coati can accomplish this odd feat because of its feet—specifically, its reversible ankle joints. Also, its long flexible tail helps the animal balance as it scampers over branches and leaps between limbs. The tail sends signals, too. Members of the same family group look for and recognize one another's tails as they travel through the trees.

BASILISK LIZARD

When frightened, the basilisk lizard has a cartoonlike ability to scamper across the surface of the water. What's the secret behind this seemingly supernatural power? When the lizard hits the water, it unrolls special long, fringy scales on its toes. The scales act like skis, spreading out the animal's light body weight. As long as the lizard's legs keep moving, it won't sink. If it does start to sink, though, it's no problem. The reptiles are excellent swimmers and can stay underwater without coming up for a breath for an astonishing half hour.

TUATARA

Talk about one of a kind. The tuatara's last living relative went extinct about 100 million years ago. Although tuataras might look like lizards, they're not! They belong to a separate group that was around during the time of the dinosaurs. This earned the tuatara the title of "living fossil." The reptile lives only in New Zealand and gets its name from a Native islander word that means "peaks on back." Both male and female animals have spiky-looking scales along their spines that stand up to attract a mate or scare off a threat.

COOL!

VIPERFISH

Some teeth aren't made for chewing. Vipers and other snakes have a reputation for swallowing their prey whole. And, despite this animal's monstrous chompers, the viperfish does the same thing, using those scary-looking teeth only to grab on to a meal. It can gobble up some pretty large animals, thanks to a special hinge in its skull. This hinge allows the viperfish to open its jaws really wide—so wide it can swallow a fish that is over half the length of its own body! It's a good thing that the viperfish's belly is super stretchy!

Macaws love palm nuts, especially ones that have already been eaten and digested by cows. Found in and around cow poop, they're nice and soft.

Each bird's coloring is unique, meaning an individual can be identified by its feathers!

MACAW

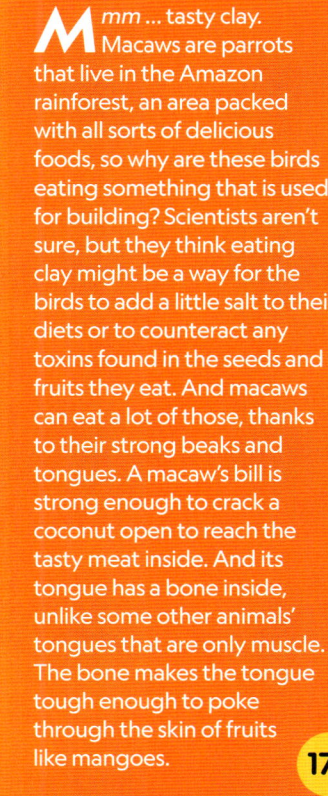

M *mm* ... tasty clay. Macaws are parrots that live in the Amazon rainforest, an area packed with all sorts of delicious foods, so why are these birds eating something that is used for building? Scientists aren't sure, but they think eating clay might be a way for the birds to add a little salt to their diets or to counteract any toxins found in the seeds and fruits they eat. And macaws can eat a lot of those, thanks to their strong beaks and tongues. A macaw's bill is strong enough to crack a coconut open to reach the tasty meat inside. And its tongue has a bone inside, unlike some other animals' tongues that are only muscle. The bone makes the tongue tough enough to poke through the skin of fruits like mangoes.

177

AFRICAN CRESTED RAT

Anything that tries to eat this extra-furry rat is in for a nasty surprise. African crested rats chew on the bark of a poisonous tree. Then they spread their saliva, which is now toxic from the bark, onto the sides of their bodies. If a rat is threatened, its poison-soaked fur fluffs up, tricking the predator into grabbing the fur with its mouth. The poison may not be enough to kill a predator—but it will probably be enough to let the rat live to see another day.

BLUE GLAUCUS

Despite its nickname—the blue angel—this sea slug is anything but innocent. The blue glaucus's funky-looking feathery parts pack a powerful punch. The slug munches on toxic animals like the Portuguese man-of-war. Stinging structures from the prey don't break down; they build up in the slug's body and are still able to cause a nasty surprise. Any animal that disturbs the blue glaucus risks receiving a jab that's super painful, or even deadly.

179

KOMODO DRAGON

Komodo dragons don't breathe fire or hoard gold. That doesn't make them any less fierce than their fictional counterparts, though. These reptiles are the world's biggest lizards and use their tongues to sense prey by tasting the air. Then they lie in wait for their meal. When an animal passes, they leap, ripping into the prey with sharp teeth. Even if the prey escapes, it's only a matter of time before this predator chows down. A Komodo dragon's spit is venomous, and the reptile will follow its meal for as long as necessary.

WhAt?!?

GOBLIN SHARK

Goblin sharks live deep in the ocean. It's dark, but that's no problem for these fish—they have special sensors that pick up faint electrical fields created by swimming prey. They also have one-of-a-kind jaws that they can move like a slingshot. Once a shark spots a fish, it goes very still, drifting as close as possible. When the prey is in reach of its jaws, the goblin shark rapidly shoots them forward to grab the fish. The shark's jaws then snap back into its mouth with its meal inside.

WEIRD ROMANCE

When it comes to attracting a mate, some animals have stranger strategies than others.

BIRD OF PARADISE

These birds would probably crush it on a TV dancing show. Male birds of paradise shimmy and shuffle back and forth in hopes of impressing a female bird.

PEACOCK SPIDER

This spider is ready to boogie! To attract a mate, a male peacock spider will dance around, waving a colorful flap that looks like a peacock's tail as it tries to get a female's attention.

GREAT CRESTED GREBE

When grebes find their match, they celebrate with an expertly choreographed dance number. Both birds dive below the water, then come up with a weed to shake from side to side.

What a wattle! Lots of birds have wattles, or long, fleshy pieces of skin that hang from their necks. But the umbrella-bird's wattle is something to see. It's covered with scaly-looking feathers that make the wattle look a little like a pine cone. Males also have a long tuft of hairy feathers that hangs over their beaks. Some people think it makes the males look like they are carrying an umbrella. Others think the bird looks just plain weird. You be the judge.

FLYING FISH

Flying fish take to the air to avoid being eaten by predators below the waves. However, once they're in the air, they can become food for birds. These animals don't really fly, but they jet out of the water at remarkable speeds of about 35 miles an hour (55 km/h), then glide through the air using special chest fins. These flights aren't little hops, either. Many kinds of flying fish can travel 650 feet (200 m) before reentering the water. That's more than the length of 10 tractor trailers!

VAMPIRE SQUID

This squid, photographed by the team at the Monterey Bay Aquarium Research Institute (MBARI), lives in the deepest, darkest parts of the ocean. If it's spotted by a predator, the bizarre-looking animal releases a cloud of liquid filled with tiny, glowing particles, then flees. The vampire squid doesn't drink blood but does feast on dead things. It collects "marine snow" to eat, which isn't snow at all, but particles of animal poop, dead plants, and other matter. As the particles fall to the bottom of the ocean, the squid captures them with the long, fleshy strands attached to the sides of its body. Gross!

No Way!!

GIANT ANTEATER

An anteater's iconic snout is probably most famous for holding its superlong tongue. It's an impressive claim to fame—its tongue is about two feet (0.6 m) long and can snag tens of thousands of insects at once. But there's more to this snout's awesomeness! Anteaters are good swimmers and use their noses like a snorkel, poking them out of the water. The snout is also related to the giant anteater's fantastic sense of smell. Scientists think the animal's sniffer is 40 times more sensitive than a human's nose.

Male seahorses hold eggs inside a kangaroolike pouch until they hatch.

PYGMY SEAHORSE

This tiny animal has a huge talent for hiding! Pygmy seahorses are the world's smallest seahorses. Many adults are less than an inch (2.5 cm) in length. That's only a little bit bigger than the average peanut! These tiny fish live around two kinds of coral found in the warm waters of the Pacific and Indian Oceans. They blend in so well that they were only discovered by chance when a scientist accidentally collected one along with a coral sample. The seahorses' lumpy bodies and pretty color make them almost invisible against the bumpy surfaces of brightly colored coral. Pygmy seahorses can be pink, purple, brown, or yellow—it all depends on the color of the coral in their neighborhood.

SPOTTED SKUNK

Is that a gymnast I see? Spotted skunks do an impressive handstand with all their limbs spread and their tails puffed out. But if you spot a skunk doing this maneuver, run! This is a clear sign they're ready to spray. Although the spotted skunk is much smaller than its bigger cousin, the striped skunk, this species is still a supersprayer. Spotted skunks can shoot their foul-smelling liquid about 10 feet (3 m) away from their rear ends!

ORANGUTAN CRAB

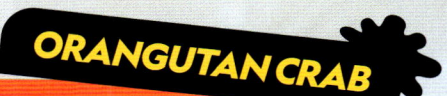

With a body only .75 inch (2 cm) wide, this little crab can be easily overlooked. You won't find orangutan crabs in trees like their ape namesakes, but they are orange and do have long, fuzzy legs like orangutans. The hairy fuzz protects them from predators and also traps tiny bits of food to eat. Orangutan crabs are members of a family of decorator crabs, which cover their bodies with bits of shells, stones, and other objects, making them less noticeable to predators.

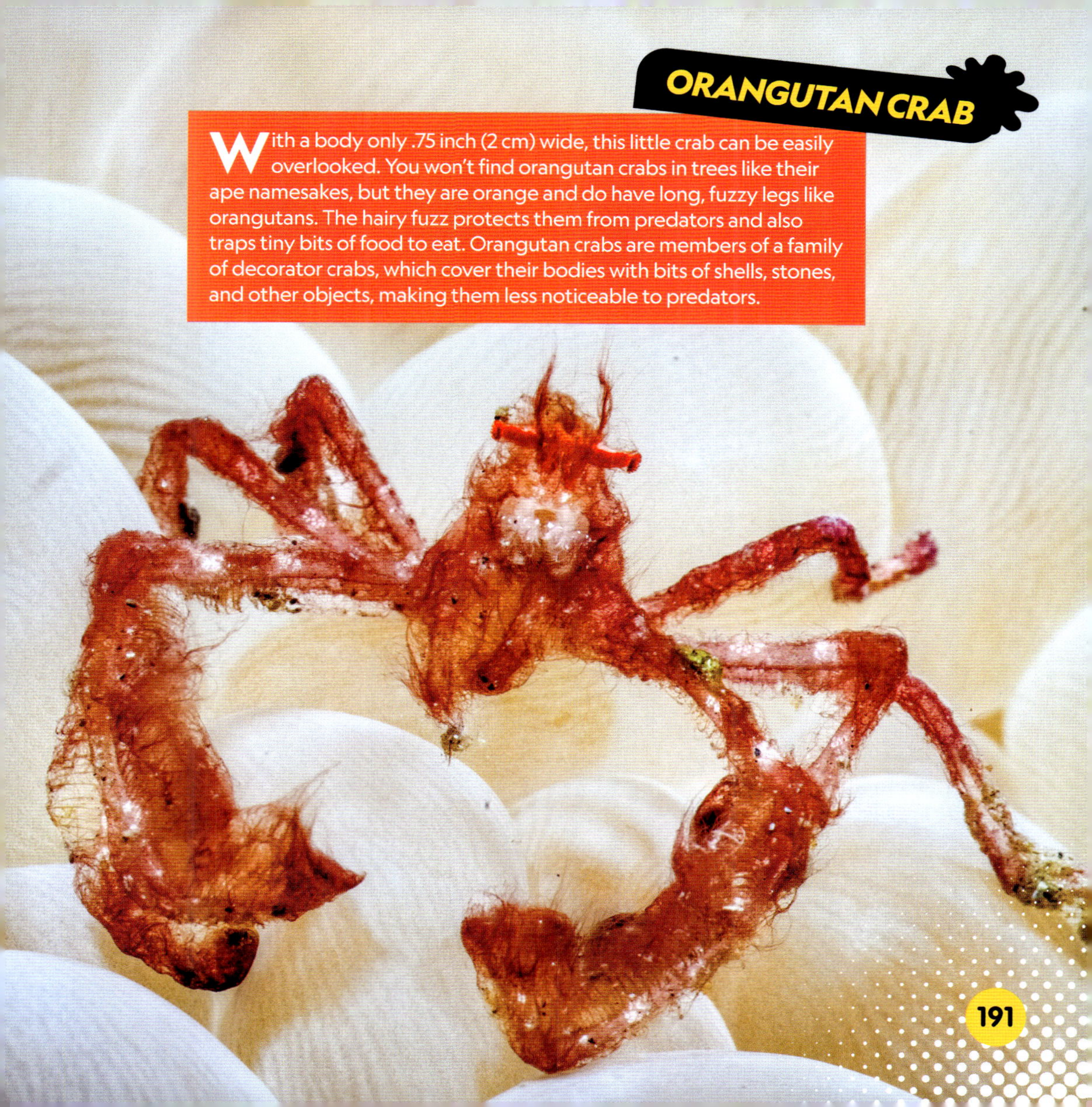

WEiRd!

SCALY-FOOT SNAIL

The scaly-foot snail may not be a weight lifter, but it does carry some serious iron. In fact, it's the only animal on Earth whose body uses iron in the formation of its body parts! Iron is found in the snail's shell and in the hard scales that cover its squishy body. This suit of armor keeps the snail safe as it scavenges for food on the ocean floor. Scientists aren't exactly sure where the iron comes from. One idea is that bacteria in its body take iron from the surrounding seawater to forge the snail's radical outerwear.

This seems like a lot of legs for one little waterbird. But only two of the legs in this picture belong to the adult jacana. The rest go with the chicks the daddy bird is nestling under his wings to protect them from lurking crocodiles. The babies' long limbs and toes make them hard to hide now, but they'll give the chicks a leg up when they get older. Jacanas have the awesome ability to walk on top of floating plants like lily pads without sinking.

PECULIAR PORTRAITS

Humans aren't the only animals that make funny faces. Check out these delightfully odd animal close-ups.

CELEBES CRESTED MACAQUE

WILD PONY

BEARDED SEAL

RABBIT

195

GIANT ISOPOD

Giant isopods look like the roly-poly bugs that can be found in rotting logs or moist soil, but they're bigger than their land-dwelling cousins—much bigger. They can grow to be almost a foot (30 cm) long. And, because they spend their lives on the deep-sea floor, scientists don't know much about them. They do know that giant isopods feed mostly on dead plants and animals that sink down from the upper layers of the ocean, but they don't eat much. One isopod that animal scientists tracked went for five years without taking a bite!

WEAVER ANT

Few animals are as crafty as these insects. Weaver ants get their name from the way they build nests. The insects work as a team to weave plant leaves and special silk strands together to create a ball-shaped nest. It's a structure fit for royalty: The colony's queen moves into the ball with her larvae, safe and protected. Weaving isn't the only talent these insects have. They're also fierce fighters. If the colony is attacked, thousands of ants will drop on the predator, spraying it with acid to drive it away.

RACCOON DOG

Don't let the dark circles around these critters' eyes fool you—raccoon dogs aren't related to raccoons. They are members of the dog family, though, and among dogs they're unique. They're the only members of the dog family known to hibernate. They also don't bark, and instead growl or whine. Known as tanuki in Japan, they often appear in Japanese folklore as merry tricksters that enjoy playing pranks on unsuspecting people walking in the woods.

KAKAPO

What's that delicious aroma? People think male kakapos smell a bit like a bakery—their feathers have a sweet or musky fragrance, sometimes smelling like honey or toasted sugar. These large birds, which are the world's heaviest parrots, are too heavy to fly, and spend most of their time in the thick plants that cover the ground of their New Zealand forest homes. The sweet smell most likely helps these parrots find each other, because they don't often have a bird's-eye view from the treetops.

COOL!

WEIRD NOSES

Brilliant beaks, super snoots, marvelous muzzles. There seems to be an endless variety of nose shapes and sizes in the animal world. Some are weirder than others.

LONG-NOSED LANTERNFLY

"Elephant fly" might be a better name for this insect. The lanternfly's "trunk" is hollow—perfect for sipping sap from tree trunks.

LONGNOSE SAWSHARK

Talk about multipurpose! This fish uses sensors on its saw-like nose to detect buried prey, then digs into the ocean bottom to uncover it.

ELEPHANT SHREW

This animal's name is half right. It's not a shrew, but surprisingly it is related to elephants. The little mammal uses its flexible snout to catch passing insects.

201

WEiRd!

FROGFISH

These fins were made for walking! Well, not quite, but frogfish do have sturdy chest fins that they use to crawl along the seafloor. These animals aren't speedy, but that's OK—they rely on elaborate camouflage to avoid being seen. And frogfish have another trick up their fishy sleeves. They are a kind of anglerfish, so they dangle a piece of skin in front of their large frog-like mouths as bait for passing prey. Any fish fooled by the bait is walking, too—right into the frogfish's trap.

KANGAROO

If you want to beat a kangaroo in a jumping contest, you'd better hop to it. Kangaroos' leaping skills all come down to some seriously strong legs. The muscles in their legs are attached to bones by superstretchy body parts called tendons. These act a little like springs, giving the animal a boost as it pushes off the ground with its big feet. Red kangaroos, the largest kind of 'roo, can clear almost 30 feet (9 m) in one jump. That's un-be-*leap*-able!

CROCODILE SKINK

This lizard's doggone weird! When the crocodile skink senses danger, it lets out a very doglike yelp to try to scare the threat away. If that doesn't work, the crocodile skink will try some other tricks, including remaining perfectly still, losing its tail, and falling over to play dead. Most predators decide that the skink is no longer a good meal and will look for prey elsewhere. The skink also uses its unique cry to defend its young. Larger animals may be too intimidated by the sounds made by this tiny croc to even come close to its eggs!

TUFTED DEER

This mammal looks ready for Halloween! Despite its peculiar vampire-like fangs, the tufted deer is an herbivore, feeding on plants that grow in central China. Unlike other deer, its canine teeth grow to be about an inch (2.5 cm) long, making them look like long, pointed fangs. The deer don't use them to chew, though, or to suck blood. Instead, male deer show them to other males as they try to determine who's boss. Male deer also make barking sounds to attract the attention of possible mates, meaning their bark is definitely worse than their bite.

ROCK ON!

Punk rockers are known for their excellent hairstyles. These birds don't have hair, but they do have fabulous feathers that are perfectly fitting for the punk scene.

HOOPOE

VICTORIA CROWNED PIGEON

COCK OF THE ROCK

207

INDEX

Boldface indicates illustrations.

INDEX

Since 1888, the National Geographic Society has funded more than 14,000 research, conservation, education, and storytelling projects around the world. National Geographic Partners distributes a portion of the funds it receives from your purchase to National Geographic Society to support programs including the conservation of animals and their habitats. To learn more, visit natgeo.com/info.

For more information, visit nationalgeographic.com, call 1-877-873-6846, or write to the following address:

National Geographic Partners, LLC
1145 17th Street NW
Washington, DC 20036-4688 U.S.A.

More for kids from National Geographic: natgeokids.com

National Geographic Kids magazine inspires children to explore their world with fun yet educational articles on animals, science, nature, and more. Using fresh storytelling and amazing photography, *Nat Geo Kids* shows kids ages 6 to 14 the fascinating truth about the world—and why they should care. natgeo.com/subscribe

For rights or permissions inquiries, please contact National Geographic Books Subsidiary Rights: bookrights@natgeo.com

Designed by Gustavo Tello

Trade paperback ISBN: 978-1-4263-7587-3
Reinforced library binding ISBN: 978-1-4263-7589-7

The publisher would like to thank Jennifer Szymanski, writer; Claire Lister, project editor; Jeremy Marshall, designer; and the packaging team at Dynamo Limited. Book team: Kathryn Williams, Emily Fego, and Avery Naughton, project editors; Lori Epstein, photo manager; Yogi Carroll, production manager; and Lauren Sciortino and David Marvin, associate designers.

Printed in China
25/LPC/3

Keep the **WEIRDNESS** coming
with **Weird But True!**

THAT'S
WEIRD!

NATIONAL GEOGRAPHIC KiDS

weird
but
true!

ANIMALS

300 outrageous
facts about
wacky wildlife

Discover
**TONS OF WILD
AND
WACKY FACTS**
about animals you
didn't even know
were weird!